John Cow

Remarks on the Manner of Fitting Boats for Ships of War and Transports Addressed to the Officers of the Royal Navy, and Royal Artillery

Elibron Classics
www.elibron.com

Elibron Classics series.

ISBN 0-543-89217-4 (paperback)
ISBN 0-543-89216-6 (hardcover)

This Elibron Classics Replica Edition is an unabridged facsimile
of the edition published in 1841 by W. Clowes & Sons; W. Cocks, London.

Ranwell del. T. Picken lith.

Day & Haghe, Lithrs to the Queen

H.M. SHIP HECLA — CAPTN W.E. PARRY — ON SHORE IN HECLA COVE,

and her Bower Anchor being laid out by her 25 ft. Cutter

August 1827

REMARKS

ON THE MANNER OF FITTING

BOATS

FOR

Ships of War and Transports:

ADDRESSED TO THE OFFICERS OF THE ROYAL NAVY,

AND ROYAL ARTILLERY.

By JOHN COW,

OF HER MAJESTY'S DOCK-YARD, WOOLWICH.

"That which is done of good-will, must, to my thinking, be accepted favourably."—Sir Walter Scott.

Second Edition.

LONDON: W. CLOWES & SONS, 14, CHARING CROSS.

W. COCKS, 18, GREEN'S END, WOOLWICH

MDCCCXLI.

TO

SIR THOMAS BYAM MARTIN, G.C.B.

ADMIRAL OF THE WHITE.

SIR,

IN availing myself of your kind permission to dedicate to you the Second Edition of this little book, I eagerly seize the opportunity it affords of repeating my most sincere thanks for the many benefits which you have been pleased to confer on me, more particularly in favouring me with your attention when I first endeavoured to explain the nature of my suggestions, and in enabling me, by your encouragement and support, to carry out those suggestions by putting them to the unfailing test of experiment.

The support with which you were thus pleased to favour me proceeded, I am persuaded, not only from the dictates of a kind disposition, but from an earnest desire on your part to foster and encourage every attempt, however humble, that had for its object the benefit of that service of which you are so distinguished a member.

I have the honour to be

SIR,

With the greatest respect,

Your most humble and obliged Servant,

JOHN COW.

Dock-Yard, Woolwich,
1*st Jan.* 1841.

CONTENTS.

INTRODUCTION.

THE favourable manner in which my little book has been received by Officers of every rank in the Naval and Artillery Services, and the many applications which have been made to me for copies since those first printed were disposed of, induce me to venture on publishing a second edition.

I hope, though perhaps at the risk of being considered presumptuous, to be able to convey useful information to those officers who may be employed in the performance of the services which are treated of in the following pages; but should I unfortunately prove unsuccessful, I can only appeal to the sentiment contained in the motto prefixed to the first edition, it being the sentiment of one well acquainted with the feelings by which the human mind is governed—"*That which is done of good-will, must, to my thinking, be accepted favourably.*"

In the exercise of my public duty, I have had the advantage of frequently conversing with experienced officers, on the inefficiency of the boats usually supplied to ships of war, and on their inapplicability to the constantly varying purposes for which they are required, the most important of which must readily occur to the mind of every one acquainted with the subject. They are,—

First.—When the boat is employed in laying out or weighing a bower anchor.

Secondly.—When she is wanted to land or embark heavy guns on a beach through a surf.

Thirdly.—When it is required to land or embark horses on a beach through a surf.

The difficulties attendant on the above services are what I have more particularly been anxious to overcome, and the result of my endeavours forms the principal part of what I here presume to submit to the Navy and Artillery.

First.—The proposition of greatest moment is certainly that of being able to lay out or weigh a bower anchor, when the weather is such that it could not be done by any mode that had before been proposed. The experience of upwards of twenty years, has placed this proposition on such secure grounds that it may with great safety be left to rest on its own merits, and the reputation it has gained. And while I make no pretension to the merit of first adopting the principle of bringing the weight of the anchor to that part of the boat best able to bear it, I trust I may justly lay claim to that of having introduced it into the Naval Service in such a state as to render it effective. I must humbly add, that the fact of three or four of our Naval officers of high rank, and of the highest character for scientific acquirements, having subsequently to the introduction of my suggestion published plans for the accomplishment of the same object, on the same principle, varying only in the details, must at least prove that the principle itself is considered to be of great importance.

Secondly.—The efficiency of the plan for landing guns on a beach through a surf, and with the same apparatus as that used for the anchor, has been fully established by experiments made at the Royal Arsenal, Woolwich, and on South Sea beach, Portsmouth; and in consequence of the favourable reports made on the results of these experiments, the plan was ordered to be adopted in the Navy by the Lords Commissioners of the Admiralty, in their order of 11th February, 1826. And I may be permitted to state, that none of the fittings proposed by the highly distinguished officers before alluded to, for laying out or weighing a bower anchor, are at all applicable to this service, namely, the landing or embarking of guns on a beach through a surf. Perhaps I may be pardoned for mentioning in this place, what I felt to be a very high compliment, that the officer appointed to command the Royal Artillery employed on the expedition against China, being the first warlike expedition from this country since the introduction of my plan, wrote to me for one of my books two days prior to his leaving Portsmouth.

Thirdly.—The difficulties attendant on getting horses on shore on a beach through a surf, have been apparent on many occasions. I am supported in this assertion by the evidence of several Naval and Artillery officers, who have done me the honour to attend to my propositions. The boat which I have constructed for this purpose possesses the convenient property of being portable, a property which cannot fail to render it particularly useful to transports when employed in the carriage of horses, where room is of so much importance.

I have considerable pleasure in stating that I do not appear with a collection of mere theoretical ideas, but with a statement of facts which, for the most part, have been the results of experiments carried on by directions from the Navy and Ordnance Boards. These experiments were made in the presence of many officers of rank and experience, copies of whose testimonials I have included in the following pages.

I have been the more induced to take this step in consequence of the Lords of the Admiralty, the Commissioners of the Navy, and the Board of Ordnance, having recognized and adopted my several propositions which are introduced in the Navy and Artillery; and it therefore appears desirable, that officers who have not had an opportunity of seeing the boats used in the manner I have proposed, should have some other means of becoming acquainted with the results of my experiments, and the mode of applying the boats. An instance of the propriety of doing so has lately occurred: one of our large steamers had occasion to lay out a bower anchor, and although her boat was fitted on my plan, there being no one on board who had seen it used for this purpose, the launch of a French frigate, which happened to be in the same port, was borrowed to effect the object.

I have given a few instances in which the

want of boats fitted as I propose has been seriously felt; or in which, if they had been used, increased convenience and safety would have been the result.

Among these instances are two which occurred in the Merchant Service; and considering it a duty to do all in my power to avert similar accidents, although not arising in the service to which I belong, I wrote to the Chairman of the Committee of Lloyd's, pointing them out, and acquainting him with what had been done in the boats in the Navy. My letter was answered by the Secretary informing me "that it was not within their (the Committee's) province to take such subjects into consideration," but recommending me "to submit the plan to the Committee of Ship-owners." I did so, and have inserted, with the other documents, a letter from their Secretary.

I am unable to say to what extent my plan has been adopted in the Merchant Service, but I have seen many boats so fitted in the River Thames.

I have taken the liberty of addressing these remarks to the officers of the Royal Navy, and Royal Artillery, because I think the subjects treated of are principally interesting to them, not doubting that the kindness and liberality by which the members of these professions are distinguished, will lead them to pardon the errors which I may have committed in my humble exertions in the cause of improvement.

Dock-Yard, Woolwich,
1st Jan. 1841.

ON THE

FITTING OF BOATS.

On the Fittings of the Launch for carrying out and weighing the Bower Anchor.

I HAD frequently heard it regretted by naval officers, that our launches were not equal to that often vitally important object of laying out a bower anchor, particularly should there be any sea on, and that the means resorted to when that service became necessary were attended with great difficulty and loss of time, it being effected either by making a raft with the top-mast, spars, &c., as practised by His Majesty's ship Newcastle, on the coast of America, in the year 1814; the carrying of the anchor between two boats; or over the stern or bow of one,—the danger attending either of which methods, should there be the least sea, must be obvious.

These considerations induced me to lay before the Honourable Commissioners of the Navy, in July, 1819, a drawing of what I then thought a new mode of carrying out an anchor, and which, from my knowledge of the capability of the boat, I was satisfied she was fully equal to.

This communication was answered by my being referred to a drawing, by which two boats were fitted for this service, and stationed at Gibraltar and Malta yards, in the years 1799 and 1804, wherein I was told that the principle of suspending an anchor under a boat's bottom was acted on, and that the Honourable Board preferred the method there shewn, to the one proposed by me.

On referring to that drawing, I certainly found that the principle was the same, but the practical execution essentially different. I have since been informed, that East India ships' boats were about the same time fitted for that object. The Malta and Gibraltar boats were disused because they were ineffective; and I am also informed that the method was discontinued in the India service for the same reason.

In the *Hampshire Telegraph* of the 20th

September, 1819, there is the following paragraph, written, I am informed, by Captain Basil Hall, R. N.

"MECHANICS.—A new and curious application of the mechanical powers has been exhibited by the natives of Porto Novo, on the coast of Coromandel, in weighing the best-bower anchor of His Majesty's ship Minden, on the Coloroon shoal, after the ship had struck and got off again, and when the ordinary process by a purchase with the launch was found to be ineffectual. The Indians formed all the spare spars, top-masts, booms, &c., into a compact body, of three or four feet in diameter, which they made fast to the buoy rope; when sixty of them, by means of slew ropes, turned the bundle of spars until the slack of the buoy rope was wound round it; when, by the judicious management of their feet, and the well-distributed weight of their bodies, they turned the spars round till the anchor was weighed; keeping it in that position whilst it was towed under the bows of the Minden, and hove up in the usual way. This anchor weighed three tons."

This article induced me again to turn my attention to the subject, and to altering the windlass so as to enable the boat not only to lay out, but to weigh a bower anchor, should it be required, as in the case of the Minden. I constructed a model, and showed it to the Comptroller of the Navy, who directed me to send it to the Navy Office, and promised he would give it his attention; he did so, and boats for between thirty and forty ships were fitted; and in consequence of the reports of their several Captains, it was finally adopted in the service, by the Honourable Navy Board's Warrant of 19th August, 1824, and by the directions of the Lords Commissioners of the Admiralty of 11th February, 1826.

Numerous instances might be adduced of the great advantages that would have been derived from ships having a boat capable of laying out or of weighing a bower anchor; and the before quoted paragraph proves, that had the Minden been so situated as not to have been able to procure assistance from the shore, she must have lost her bower anchor and cable. A distinguished officer in the Navy told me, that a ship which he commanded was lost because she had not a boat that could lay out a bower anchor; and had his boat been fitted on my principle, she would have been saved.

In Captain E. Brenton's *Naval History*, it is stated, when relating the loss of His Majesty's ship Minerve, that "Captain Brenton, aware that strong and decided measures were necessary, and that the launch of a frigate was not calculated to carry out a bower anchor, immediately despatched his boats armed to cut out a vessel from under the batteries, of sufficient capacity for the purpose. A boat proceeded under a heavy fire of round, grape, and musketry, and from her position close to the batteries, cut out a lugger of fifty tons, laden with stone for the works, and towed her off to the ship. Before the bower anchor could be placed in this vessel, it was necessary to clear her of her cargo."

I may also mention the loss of the Mulgrave Castle, East India free-trader, at the Cape of Good Hope, as appears from the Captain's deposition, published in the *British Press* of the 25th of November, 1825; which ship would certainly have been saved, had her own boat been capable of laying out a bower anchor. But the most distressing and heart-rending case is published in *Lloyd's List* of the 15th of

April, 1824:—"A tremendous gale was experienced at Alvarado, on the 11th and 12th of February. A launch belonging to the United States' schooner Shark was sent to assist an English schooner, that run on shore during the gale, and got her off; when the boatswain and sixteen men were left to weigh an anchor, but the boat was upset in the gale, and all, except one man, were drowned."

It is considered that sufficient has been advanced, to prove the advantages of a boat so fitted as to be capable of laying out a bower anchor; and it now becomes necessary to describe the manner of doing it, and to prove that every ship has a boat more than doubly equal to that service, with perfect safety.

The principle is to bring the weight of the anchor on that part of the boat most capable of bearing it; and to use a purchase in the boat equal to the heaving up of any weight that she can sustain.

To accomplish which, two solid pieces of wood, about ten inches wide, are let through the bottom, and firmly secured to the timbers; the centre of each piece is sixteen inches from the middle line of the keel, and as nearly as possible in the middle of the boat longitudinally; in each of these pieces there is a hole, large enough to take a nine-inch rope; over these holes, water-tight trunks (turned to the form of columns) are placed, and secured at the upper part by pieces of plank, which receive the heads of the columns. These pieces of plank are secured to the main thwart, and to the thwart abaft it, by forelocked bolts, which prevent the possibility of the trunks moving; and when they are in their places, the water flows up them to its level. It must be observed, that the holes through the boat's bottom are half an inch less than the holes in the trunks, so that should it be required to remove the trunks whilst the boat is in the water, plugs made of fir, with an eye-bolt driven through them, may be dropped down the trunks into the hole in the bottom; and by a blow with the end of a boat-hook staff, they are fastened, so that the trunks may be removed, and no water come into the boat. Also should it be required to put the trunks in their places whilst the boat is in the water, it will be only necessary to make a small rope fast to the eye-bolt in the plug, reeve it through the trunks, and put them in their places; bring the small rope to the windlass, and the plugs will come up the trunks, and, as in removing them, no water will come into the boat. The object in fitting the trunks in this manner is, that when the boat is not wanted for the particular service of the anchor or guns, the trunks are stowed away, and she is in precisely the same state as other boats, and no impediment offered to the stowage of water casks or any other articles.

The windlass or purchase for heaving up the anchor or guns, is composed of two parts, and when in use it is joined in the middle by an iron gudgeon; it has likewise gudgeons at each end, working in brass rhodings. The middle one works on a strong iron stanchion, which is placed on the keelson of the boat, consequently the strain is divided between the sides and the keel; and the windlass being made of two pieces of oak timber, from four to five feet long, and from seven to eleven inches in diameter each (according to the breadth and size of the boat), and being well-secured with iron hoops, it is impossible to break or bend it: therefore, it may with truth be said it is equal to the heaving up of any weight with which the boat can swim.

There is nothing complicated in the fitting, consequently it is free from the objection which is always attached to machinery on board ships, as it is worked with iron levers or handspikes, which every seaman is in the frequent habit of using. Latterly, however, it has been considered that the windlass being made in two pieces is not sufficiently strong, and that the middle iron gudgeon is liable to break; therefore I have introduced windlasses made of one piece, hooped in the middle, and working on a strong iron crutch and stantion. This method increases the friction, and is not so readily stowed away; but perhaps it is the safest.

It now becomes proper to state the tonnage of boats belonging to ships of war, and the weight of their anchors and cables, when it will be evident that there is no doubt of their capability.

	Tons.
A 74-gun ship's launch measures.	20
A frigate's ditto	15
A 32-feet barge, or pinnace, (usually issued to sloops)	8
A 26-feet yawl, (usually issued to brigs) . .	5½

	Tons.	cwts.	qrs.	lbs.
The anchor for a 74-gun ship, weighs .	3	16	0	0
Cable, 22 inch . .	5	2	1	14
Anchor and cable .	8	18	1	14
The anchor for a large frigate, weighs .	2	8	0	0
Cable, 18 inch . .	3	8	1	14
Anchor and cable .	5	16	1	14
The anchor for a small frigate, weighs .	1	9	2	0
Cable, 14½ inch .	2	4	2	12
Anchor and cable .	3	14	0	12
The anchor for a sloop, weighs . .	1	5	0	0
Cable, 14 inch . .	2	1	2	20
Anchor and cable .	3	6	2	20
The anchor for a brig, weighs . .	1	0	0	0
Cable, 13½ inch. .	1	18	3	0
Anchor and cable .	2	18	3	0

By the above statement, it will be immediately seen, that should the boats be considered sufficiently laden with only half their tonnage, still there will be an excess of capability in the boats whilst carrying the anchor and cable. As a further proof of which, a bower anchor was suspended from the bottom, and at the same time *two* bower cables were coiled into the Seringapatam's launch, which she carried well;

and they did not, to any great extent, impede her passing through the water, for she was towed by two pinnaces, at the rate of one knot per hour, against a flood tide in the Tagus. And in order to ascertain what weight pinnaces and smaller boats would carry, six tons of iron ballast were put into a thirty-two feet pinnace, and she had then fourteen and a half inches freeboard; also, five tons of iron ballast were put into a twenty-six feet yawl, and she had seventeen inches freeboard. It should, likewise, be remembered, that the anchor by being immersed in the water loses nearly one-eighth of its weight, so that there can be no doubt of the capability of the boat.

It may be proper here to state, that when the above was written, chain cables were not so universally used as at present; therefore the weight is very materially increased, and the difficulty of performing this service proportionally greater. Consequently recourse must be had to assistance from the other boats to carry the chain; but at the same time it must be evident, that the increased difficulty applies to any other mode of laying out an anchor with a chain cable.

In the establishment of cables for all ships in the Navy, hemp ones are included; and, perhaps, time and circumstances may admit of bending one of these, which would in a very great degree lessen the difficulty.

It is submitted, but with great deference, that the best way of slinging and carrying the anchor is to make the bight of a rope (the size of course will be governed by the weight, but the two parts of a seven-inch rope will carry any anchor) fast to one of the arms or flukes; also, a rope to the ring; to lower the anchor from the cat-head, sufficiently to prevent the boat striking on it; then reeve the two parts of the fluke ropes through the trunks, bring them to the windlass in the boat, and heave the pee as close up as circumstances may require; bring the ring rope over the stern, (or bow, if it should be preferred,) and a small luff tackle will easily bring the shank to a line parallel with the keel; or it may be carried with the arms horizontally, and the stock perpendicularly; but the length of the shank not being so much as half the length of the boat, it is considered that it could not thus be got so close to the bottom—the manner of doing it must of course be left to the discretion of the officer conducting the service, and perhaps local circumstances may determine which way may be best.

It has been asserted, that when it is required to lay an anchor out, the ship will be aground, and there would not be sufficient water to carry it under the boat. A short view of the draught of water of the ship, the spread of the anchor, &c., will, I presume, show the fallacy of this assertion.

A 74-gun ship draws forward . . .			23 ft.	0 in.
The spread of the anchor, .	8 ft.	10 in.		
The draught of water of the boat, with the anchor suspended .	3	2		
Admitting that it could not be hove up close by	2	0		
Making altogether			14	0
Difference . . .			9 ft.	0 in.

Still the anchor would be nine feet from the ground; and it is considered, that should there be so much sea as to cause the boat to fall nine feet, it would be impracticable to lay an anchor out by any means that could be devised; and it must always be remembered, that if, from circumstances, it should be inadvisable to put this method into practice, there are still the same resources as heretofore.

It is also submitted, and with the same deference, that the best way to pick up an anchor, is to sweep for it with the bight of a hawser. The idea of weighing it by the buoy rope appears to be exploded in the Naval service, and consequently the size of the buoy ropes has been considerably diminished; for although they may be sufficiently strong, still it would not be prudent to depend on the buoy rope, provided it were possible to sweep the anchor; for should it be chafed, or from any cause break, most probably the anchor would be lost altogether. But by keeping the buoy rope tight, you have a mark whereby you can sweep for the anchor, and when this is accomplished, a shackle or jewel may be sent down to jam the two parts; the ends may then be rove through the trunks already described, brought to the windlass and hove up. Twelve men weighed, with great ease, an anchor weighing forty-six hundred weight, after it had been down three weeks, during which time there was a heavy gale for two days. Twenty-four men can be placed at the windlass should it be necessary.

It is likewise submitted that circumstances may occur, where it would be useful in unmooring or mooring a ship; such as in crowded or narrow roadsteads. And, also, should it happen that one ship in a roadstead parts, and brings up near the shore with her remaining anchor, and after the gale has abated it becomes desirable to give the ship in distress an anchor and cable from one which rode the gale out, it would be most likely that the ship in safety would be to windward; in that case, the anchor might be dropped in any required spot, and the end of the cable given to the ship in distress.

Admitting that through the whole course of a ship's services it may not be required to use the apparatus, (and it is much to be desired that it may not,) still there is a positive and every-day felt advantage in this manner of fitting boats, arising from the improvement it admits of being made in their form.

The present launches have a broad and deep transom, and are very full abaft, which must impede their rowing and sailing, and also adds to their weight, which form is intended to enable them to carry an anchor over their stern, (which even now they will not do, should there be the least sea on,) but as the principle is established of weighing anchors by the midships of the boat, the necessity for their fulness abaft does not exist; consequently, a better formed boat may be constructed, without at all interfering with her internal stowage, or general utility.

At the suggestion of Rear Admiral Sir Thomas Hardy, the Honourable Navy Board ordered launches to be built in this dock-yard for the Ganges, Barham, and Hussar, and one at Portsmouth for the Asia, on the altered principle; and as far as can at present be ascertained, they are, according to the opinion of naval men, much improved boats.

At Rio Janeiro it is usual to employ all the boats of the fleet to tow ships out of the harbour, and when the vessel is at the harbour's mouth, the boats are cast off, and generally vie with each other in pulling on their return to their respective ships; on those occasions, the Ganges' launch often held a good tug with a thirty-two feet cutter, but the other launches never could come near her.

Should it be observed that the expense of building boats is increased by these fittings, it may be answered, that if only one line-of-battle ship's anchor and cable be saved in seven years, it will more than counterbalance the expense of fitting the whole of the ships in commission for that period, in time of peace; but the cost of an anchor and cable is of little importance, compared to the insecurity a ship is placed in when in want of them, particularly on foreign stations; independent of its "affording the means of saving a ship from wreck, when no other means could be obtained."

I must here take the liberty of inserting an extract from Captain Parry's last voyage towards the North Pole:—"I cannot here omit to mention the invaluable advantage derived on this occasion from one of our cutters (a twenty-five feet boat) having been fitted on Mr. Cow's ingenious principle, for weighing anchors in the centre, instead of the extremity of the boat. By this beautiful contrivance, six men could weigh the Hecla's bower anchor of thirty hundred weight with ease, and transport it any distance with safety; indeed but for this facility, added to that afforded by Phillips's capstan, the Hecla's reduced crew would probably have been unable to haul her off the ground on this occasion." In addition to which, Lieutenant (late Captain) Foster, who commanded the Hecla during the absence of Captain Parry, and on the occasion above alluded to, told me, "Had it not been that our boat was fitted on your plan, we should have lost the Hecla."

On Landing and Embarking Heavy Ordnance.

In expeditions to an enemy's coast, the heavy artillery attached to the army is conveyed in vessels appointed expressly for that service, which are called Ordnance Transports; and when required to land the guns, it is generally done with their boats, under the directions of the officers of the artillery, and with the assistance of the artillery-men.

Should there be a wharf, jetty, or projecting bank, no difficulty would arise in accomplishing it, as a purchase could very soon be established for hoisting them out. Many ingenious and effective methods for doing so are practised in the Royal Military Repository at Woolwich.

The present object is to suggest a method of landing large guns on a beach, where there is so much surf that it would be certain destruction to a heavily laden boat to approach near enough to parbuckle them out, for she must take the ground,—as, from the great weight of the gun, it would not be prudent, or indeed practicable, to do it while she is afloat.

I understand that in these cases it is usual to erect a temporary wharf; but to make that sufficiently strong to bear the weight of a thirty-two or twenty-four pounder gun must of necessity be a work of some time, and the materials requisite would occupy a considerable space, and are not always to be obtained.

Should it not be possible to accomplish that method, a pair of sheers, or a triangle, has been erected over the boat, and the gun hoisted out by a purchase from it; but it must be evident that it cannot be done in this way, should there be any surf or sea to give the boat much motion, as she would be continually striking against the sheers or triangle, and thereby liable to the danger of upsetting them.

So much difficulty always attends the landing of the heavy ordnance attached to an army, that I believe it is seldom attempted until some harbour or convenient spot is acquired to do it in. The possible distance of such harbour from the scene of operations, and the consequent loss of time and labour in bringing the guns up, are so obvious, that they need no comment.

Independent of the artillery attached to an

army, it has very frequently occurred that ships of war have landed their lower-deck guns, to erect breaching batteries; in doing which they are of course subject to the same difficulties as the transports, only they have generally larger boats, and perhaps more men disposable for that service. And yet, with these advantages, and with the characteristic zeal and ability of British seamen, they cannot at all times overcome the difficulties of landing heavy guns through a surf. Every effort was made by Captain (now Sir Josias) Rowley, when in command of a squadron in the attack on Leghorn, in December, 1813.

A part of the Italian levy, with marines and seamen of the squadron, were landed, but it was not possible to get the guns on shore, which was the sole cause of the failure of the enterprise.

A circumstance occurred on the coast of Africa, which was related to me, at different times, by two officers who were employed on the service, and which, in my humble opinion, fully confirms the difficulty of landing heavy guns by the ordinary means. It is as follows:—Two transports, the Anne and Turners, were laden at Woolwich, one on the 31st of December, 1823, and the other on the 12th of January, 1824, with ordnance stores, and had thirty-two, twenty-four, eighteen, twelve, nine, and six pounder iron guns on board, consigned to the coast of Africa. On the arrival of those ships, the usual method of landing the guns was resorted to, (with assistance from the Owen Glendower frigate,) which was by means of a raft, as the surf ran about forty fathoms from the shore; but so much difficulty, risk of losing the guns, and danger to the people, arose in getting the small guns on shore, that they would not attempt to land the large guns, and the ships were obliged to take them to the West Indies.

I am not in possession of the details of the business, so as to enable me to state the number and size of the guns which they could not land; but I am satisfied as to its correctness, and perhaps I may be permitted to state, that the officers from whom I derived my information were decidedly of opinion, that had the method about to be described been put in practice, there would not have been any doubt of its being accomplished. It must be presumed that these guns were necessary for the defence of our settlements on the coast, therefore the disappointment and probable injury to the service must be evident.

The embarkation of guns is attended with the same, and on some occasions with much greater difficulties than have been described in landing them. In a great degree to lessen the difficulty, and entirely obviate the danger, both to men and boats, is the purport of the present proposition.

Repeated trials with the anchor had fully convinced me of the power of the windlass, and the unimportant effect which the weight of the anchor had upon the boat; therefore, I was led to consider, that, without any additional fittings, the apparatus might be made available to the landing and embarking of heavy guns, in the following manner:—

The trunks and windlass are put in their places in the boat, as before described for the anchor. The gun is thus prepared for landing:—a pair of gun-slings is put on in the usual way; a single rope (which I shall call the heaving-up rope) about seven inches in circumference, is made fast round it, exactly in the middle, so that the gun shall balance. It is then removed from its carriage, (should it be mounted,) and placed on a cradle or sledge. Should this sledge not be previously prepared, it may be made of any piece of timber that can be procured; if on board a ship of war, the spare half anchor stock would answer perfectly well, by being cut asunder, reversed, and the two small ends brought and bolted together. The upper part should be hollowed out to the form of the gun, so that it may lie more steadily on it; the bolts which fasten the two parts of the sledge would serve to pass the lashings round to secure the gun on it. That end of the sledge which is intended to go foremost, should be rounded away, so as not to offer any obstruction, and to rise over any impediment on the ground that it may meet with. A strong rope is then fastened under the muzzle of the gun, for hauling it on shore: it should be particularly observed to place the rope *under* the muzzle; for on hauling it on shore, the people are necessarily more elevated than the gun, and consequently the rope by being so placed tends to lift it, and in a great measure to prevent its ploughing into the ground.

'Small lines are to be fastened to the muzzle and breech, to keep the gun steady; those lines are to be marked, so as to indicate when the gun is in a fore and aft direction. It must be particularly observed that these lines should be small, only just sufficient to steady the gun, for this reason:—should it happen from any cause, which cannot be foreseen, that the rope which heaves the gun up, slips or gives way, when the steadying lines are made fast, the gun would swing out, and its whole weight would be brought to the gunwale, when, if those lines are so strong as to sustain the weight, it would nearly upset the boat; therefore, to prevent the possibility of an accident, pieces of about an inch and a half line would be best for the purpose.

The muzzle and breech should be padded with swabs or mats, to prevent injury to the bottom of the boat.

The gun, thus prepared, should be hoisted out by the gun-slings in the usual way, and lowered sufficiently into the water; the heaving-up rope is then to be rove through one of the trunks, and brought to the windlass, and when the weight of the gun is on the boat, it should be disengaged from the tackle which hoisted it out, and hove as close as possible up to the bottom of the boat.

In order to relieve the windlass, and to enable the second gun to be hove up, a piece of timber, or three capstan bars lashed together, should be placed across the boat; a stopper should then be made fast to the heaving-up rope, and fastened to the bars; the rope round the windlass should be slackened up, and the first gun will be suspended from the capstan bars.

A mouse or knot should be worked in that part of the heaving-up rope which is nearest the windlass, to prevent the stopper from slipping.

The lines for steadying the gun are made fast to the gunwale of the boat.

A seventy-four gun ship's launch, with one iron twenty-four pounder gun suspended, heels only eight inches; therefore it is in perfect safety.

The second gun should be prepared and hove up in the same manner as the first, and left suspended from the windlass, and the boat is thereby brought on an even keel.

When both guns are secured, the carriages may be placed in the boat, with whatever stores are necessary, care being taken to avoid impeding the rowers, should the wind be contrary.

The guns by being suspended under the bottom of the boat certainly form an impediment to her progress through the water, but not so much as might be expected, as was proved by the experiment I was permitted to make in the Thames near the Royal Arsenal, on the 18th of July, 1822.

On approaching the shore, great care must be taken to bring the boat to an anchor in proper time, without the surf; for should the gun be allowed to touch the ground whilst suspended, it would inevitably stave her, and the consequences might be very serious; therefore, the depth of water must determine the distance.

Should there be a sufficient number of men on shore to haul the gun up, the only communication necessary would be a small line, by which the large hauling rope might be conveyed; then a proper number of men or horses being put to the rope, the gun could be drawn up the beach. Forty men have drawn an iron twenty-four pounder gun up a rough causeway.

Should the operation be carried on where there is a rise and fall of tide, and should the guns not be immediately wanted, it will be only necessary to take the boat in at high-water, and drop the gun until the tide leaves it.

When embarking the gun, the boat must be anchored without the surf, with an anchor sufficiently heavy to withstand the weight of the gun, and the impediments it may meet with in being hove off.

The gun should be prepared in precisely the same manner as for landing, and it should be placed as near the water's edge as possible; the rope for heaving it off should be conveyed to the stern of the boat; and should there not be a sufficient number of men to haul it off by hand, it must be brought to the windlass and hove off by that purchase. It is here proper to notice, that the windlass is *only* secured for a perpendicular strain, and if this rope is brought to it direct from the stern of the boat, the strain would be horizontal; to obviate which, an eye-bolt is driven through the boat's bottom, directly under the windlass, to which a leading block should be hooked; the rope must be rove through it, and then brought up to the windlass; by doing so, the strain is kept perpendicular.

Twenty-nine men have hauled by hand a twenty-four pounder gun from the water's edge, to the stern of a boat, moored about thirty fathoms from the shore.

The heaving-up rope and the steadying lines are to be attached to the heaving-off rope, so that they may be got at when the gun is under the stern.

When the gun is as close as possible to the stern of the boat, the heaving-up rope should be rove through the trunk, and brought to the windlass; the cable should be slackened until the boat is directly over the gun, when it should be hove up, and secured as for landing.

The second gun is got off, and secured in the same manner, when the boat may proceed to the ship and deliver them.

The advantages presumed to be gained by

this mode of landing and embarking guns, are—

Expedition;

Accomplishing the object without endangering the boat or men;

Performing such service at a time when it could not be done by any other means formerly in use;

After landing guns, the boat is instantly ready for any other service, by her being kept afloat beyond the surf.

I solicited and obtained permission from the Honourable Board of Ordnance, for a committee of Colonels and Field-Officers of Artillery to examine a model I had constructed, and to report thereon. The committee suggested that a boat should be fitted under my inspection, and a trial made on the Thames. This suggestion was attended to by the Honourable Board; and I was directed to superintend the fitting of the boat, and was informed that orders were given for my being permitted to conduct any experiments that might be made.

The boat was fitted, and two iron twenty-four pounder guns were landed, and re-embarked, in the presence of Lieut.-General Ramsay, the Commandant of the garrison, and a full committee of Colonels and Field-Officers of Artillery, who expressed themselves perfectly satisfied, and recommended that it should become an exercise of the regiment, which His Grace the Master-General and Board have been pleased to order.

I must here beg permission to acknowledge my grateful feelings for the very great attention and urbanity which I received from the Lieut.-General, the Brigade Major, and the different officers of the Artillery; who most willingly granted me whatever assistance I required, and appeared anxious that I should succeed in my endeavours to bring forward what they were pleased to term "an improvement in the Artillery service."

I also solicited the Honourable Navy Board to draw the attention of the Lords Commissioners of the Admiralty to the subject, considering it might be useful to the Royal Marine Artillery. Their Lordships were pleased to direct Lieut.-Colonel Sir Richard Williams, the Commandant, to inspect and report on it; and in consequence of the Lieut.-Colonel's report, their lordships ordered a boat to be fitted and sent to Portsmouth, for the use of that corps; and on the 7th of August, 1827, a trial was made on Southsea beach, for landing two eighteen-pounder guns from a frigate's launch, in presence of His Royal Highness the Lord High Admiral and several naval officers.

The two guns were hoisted up under the boat's bottom by means of her windlass, and the carriages were placed in the launch, between the thwarts. The launch was veered in towards the shore, and checked at about fifty yards from the beach, when the guns were lowered down from the bottom of the boat; and upon the presumption of a surf, and a tide that would render it impossible to convey a line to the shore, a brass 5½-inch mortar was placed in the stern sheets, and a shell (to which a line was attached) was projected from it, with a charge of two ounces of powder. The carriage and gun were hauled on shore, the gun mounted and fired; from the time of firing the mortar, to the discharge of the gun, fourteen minutes only elapsed.

A circumstance has occurred in which the having a boat fitted in the manner I have proposed was found to be exceedingly useful; and as the application to this particular service, viz., the picking up of guns which had been thrown overboard, was not contemplated in the first instance, I trust it will be considered right that I should mention it, and perhaps I cannot do better than by inserting a copy of a letter I addressed to the Editor of the *United Service Journal* shortly after its occurrence, and which the Editor was pleased to insert in that valuable publication.

" *Dock-Yard, Woolwich, Sept. 14th*, 1830.

" MR. EDITOR,

" Convinced that your valuable journal is open to every endeavour that has for its object the benefit of His Majesty's Service, and being further encouraged by the notice which, in the number for November last, you took of the book I was induced to publish explanatory of methods proposed for fitting boats for weighing anchors, landing guns, &c., induces me to hope, that I shall not be considered as improperly intruding myself on your notice, by stating the manner in which the guns were recovered that had been thrown overboard from the Barham, when that ship got on shore in the West Indies, in April 1829.

"The Barham having been paid off at this port, afforded me an opportunity of acquiring the following information, which I received from an officer who was in the boat when the service was performed.

"After the ship got afloat and into Curacoa, it became desirable to pick up the guns, and also a bower anchor which had been left where the ship grounded. The Barham's launch being under repair, the Admiral ordered the Druid to supply her launch, which was fitted with my apparatus; this launch, accompanied by a schooner, was sent to effect this object. A diver went down and rove a chain through the loop on the cascable of the gun, one part of the chain was brought over the gunwale and made fast to the ring-bolt in the keelson, the other part was taken up the trunk to the windlass, and the gun was hove up by the windlass, close to the bottom of the boat, hanging perpendicularly; a bight of a rope was passed over the bows, which caught the muzzle of the gun, this rope was taken aft over the transom, and with the assistance of a luff tackle, the gun was brought to an horizontal position; they stoppered the heaving-up chain, and by that means relieved the windlass, and then hove the second gun up by the chain passing through the *other* trunk to the windlass. Two guns were got up in about thirty minutes, and were carried under the bottom and delivered to the schooner, which was lying between two and three miles off.

"In this manner they picked up thirty-seven 32-pounder guns, *viz.*, twenty-two of 48 cwt., and fifteen of 56 cwt.; and afterwards weighed and carried the same distance the Barham's bower anchor of nearly 70 cwt., and I must beg to be permitted to remark that this was done by a *frigate's launch.*

"I will not presume to say that some other method would not have been found to recover the anchor and guns, for I believe the resources of our Naval Officers to be equal to the emergency which may call them forth, but at all events the launch being so fitted was found most useful.

"I cannot conclude without expressing my admiration, as far as I am capable of judging, of the excellent and seaman-like manner in which this service must have been performed by the Officers and people of the Barham.

"I have the honour to be, &c.

"John Cow."

On Landing and Embarking Field Guns mounted on their Carriages.

ENCOURAGED and stimulated by the manner in which my proposition for landing heavy guns was received, and having both read and heard of the difficulty of landing mounted field guns, I was induced to endeavour to extend the principle to the effecting of that object.

When it is required to land field guns, they are generally dismounted, the wheels taken from their carriages, and the gun, carriage, and its wheels removed from the boat separately; and when on shore, the wheels are put on and the gun remounted.

It is true, from the high state of perfection to which our Artillerists have arrived, it is the work of but very little time to mount a field gun; but should the landing be effected in the face of the enemy, the shortest period may be of the greatest consequence.

It has on many occasions been considered as very important, to be able to get a gun on shore ready for immediate use.

In the expedition to Egypt, a frigate's launch was fitted for landing a six-pounder, and it was done in the following manner, as described to me by the late Admiral Sir Benjamin Hallowell :—

The foremost thwarts were unshipped, and skids laid in a declined direction, from the bow to the bottom of the boat, on which skids the wheels of the carriages were placed ; there were also skids from the bow to the shore; by this method the gun was kept as low in the boat as possible, and ran up the inclined plane of the skids, and then over the bow.

In America, at the close of the last war, a line of-battle ship's launch was fitted with a platform on the gunwales, on which platform a nine-pounder gun was placed, and ran out on skids from the bow, as in the former case.

It is presumed, that by either of these methods it cannot be accomplished, or at least without imminent danger, should the beach be in such a state as to preclude a near approach.

Also in the latter case, the gun, by being placed so high, would make it impossible to carry sail on the boat, and would render her unsafe should she have much motion.

Likewise, the gun is much exposed, and liable to be rendered useless, before it reaches the shore, by the shot of the enemy.

In either of the before-mentioned cases, it will be observed, that the six or nine-pounder guns engaged, and nearly occupied the launch of a frigate or a line-of-battle ship, besides rendering more than half their oars useless; and most likely they would require fortuitous aid, as was evinced at the landing in Egypt, where "fourteen launches, each having a field-piece and her own carronade, twenty-five seamen, and eight artillerymen, besides the boat's crew," required "fourteen rowing boats to tow them."

On several expeditions to the enemy's coast, the landing has been delayed by the state of the weather; and although it might have been possible to get the troops on shore, it was not practicable to land the guns, therefore it was not attempted; the ill effect of such delay may easily be conceived.

The expedition to the Helder, in 1799, clearly

exemplifies the great evil of not being able to ensure the landing of the guns at the same time as the troops.

On that expedition a part of the troops were landed, but without any guns; the surf got up on the beach, and for two days no further disembarkation could be effected: "those troops were attacked, but repulsed their assailants, although without artillery or cavalry;" but had the enemy been in force in the neighbourhood, the consequences might not only have been very serious to those who were on shore, but might have caused the total failure of the enterprise.

These considerations induced me to solicit permission to try whether a mounted field gun could not be got on shore by a different method from those above described; and instead of occupying the largest boat for a six-pounder, to effect the landing of a twelve-pounder by a twenty-five feet cutter, and to ensure its being landed at the same time as the troops.

I must admit it was an extreme case, but the object in view was, to ascertain with what facilities it might done in a river, so as to form an opinion of what would be required in more exposed situations.

The Lords Commissioners of the Admiralty were pleased to direct the Captains of the ships fitting out at this port, in conjunction with a committee of Artillery Officers, to witness, and report on the experiment I had proposed to make.

A twenty-five feet cutter was fitted with the trunks, and a temporary windlass put into her.

A brass twelve-pounder gun (of 18 cwt.), with the limbers, magazine, shot-rammer, &c., was placed on the wharf at the west end of the Arsenal. The gun was first disengaged from the limbers, and two ropes made fast to it round the axle, one on each side between the wheel and the carriage, and so placed that the gun and carriage balanced. A suitable rope for hauling it on shore was made fast to the trail, as were also two lines for steadying it when under the bottom of the boat; it was then lowered from the wharf by the crane, the heaving-up ropes were rove through the trunks, brought to the windlass, and the gun and carriage were hove up as close to the keel as possible; the steadying lines (previously marked) were made fast, and the gun was kept by them in a fore and aft direction ; the wheels were taken from the limbers, and, with the shafts, were placed in the stern sheets of the boat ; the magazine, shot, &c., were put in midships under the thwarts. The boat was shoved off, and rowed with eight men the whole length of the Arsenal (nearly half a mile), and brought to an anchor about twenty fathoms from the shore. The gun was lowered down, and the rope which was fast to the trail was given to twelve artillerymen, who ran the gun up the causeway with the greatest possible ease. The boat was then put on shore, the limbers, &c., handed out, and the gun was limbered up, and ready for advancing in less than five minutes.

By using proper precaution, the gun might have been loaded previously to its being put into the water, and it might have been fired the moment it reached the shore.

The gun is prepared for re-embarkation in the same manner as for landing, and it is accomplished in a similar way to that which has been described for heavy guns.

Powder barrels and cases are water-tight, and so are the ammunition-boxes attached to the gun; but to prevent the possibility of an accident, a strong box should be made, sufficiently large to contain two barrels, which box might be drawn through the surf at the same time as the gun.

I have been favoured with a copy of the Naval Officers' report, also that of the Committee of Artillery Officers, which are annexed; and as it details many advantages likely to be gained by this method, I shall only take the liberty of mentioning one, which, it is humbly submitted, may be of very great importance.

Should the landing of the gun be effected by a pinnace, yawl, or any boat fitted for a carronade, it does not cause any impediment to fighting it; so that the carronade in the boat might be used either for assisting to clear the beach before the gun is landed, or for covering it until it is ready to advance.

On the Construction of a Portable Boat for Landing and Embarking the Horses attached to a Field Gun, or Cavalry of any description, on a Beach and through a Surf.

ONE of the most pleasing circumstances resulting from my endeavours to introduce improvements in boats, is that of giving me an opportunity of conversing with officers of rank and experience, and thereby obtaining the necessary information for carrying on the several public services, and what alterations would be desirable in them could they be effected; a species of information which, from my station in life, I could have no means of obtaining by my own observation. It was during a conversation with Captain M'Kinley, R.N. (who has frequently done me the honour to inspect my propositions), that he mentioned the difficulties that usually arise in landing or embarking horses on a beach, should there be any surf; and he also particularly stated what took place at the Helder, in 1799, when, after the General (Sir Ralph Abercrombie) had landed, it was found impracticable for a considerable time to get a horse on shore for his use, which he was exceedingly anxious to effect; it being of the greatest importance that he should be mounted. It was at length accomplished by a Captain in the Navy, who volunteered his services, and mounted the horse, and swam him on shore through a great surf, at the imminent risk of himself and horse. Also when the experiment was made at the Royal Arsenal, of landing a field gun mounted on its carriage, and when the limbers were got on shore and the gun limbered up, an Artillery Officer, one of the Committee, addressed me, and said,—"You have got the gun on shore, and it is ready for moving forward, but the same difficulty of weather and surf would prevent your landing the horses attached to the gun, and therefore prevent its advancing with the celerity that might be wished." These conversations forcibly impressed on my mind the necessity of some plan to accomplish this object, and to be enabled to secure the landing, under the same circumstances of weather and surf, not only the gun but the horses, when the beach is in such a state that men could be landed. I, in consequence, set about making the most diligent inquiries, not only at the model-room of the Military Repository at this place, but also of all the officers, old soldiers and sailors, with whom I could communicate, as to whether there was any established method of performing this service; and if not, what means were generally resorted to.

Although there are models at the Repository, showing the manner in which horses are hoisted in or out of transports when alongside of a wharf, I could not discover any which applied to the landing of them on a beach; and all the information I could collect was, that they were either placed in flat-bottomed boats, or that two long-boats from the transports were lashed together, and a platform laid on them, with stanchions and a guard rope to prevent the horses from going overboard. It must be evident that should the surf extend any distance from the shore, it would be utterly impossible to accomplish the object by either of the methods here mentioned; and the only resource left would be, to get the transports as near as possible to the beach, to hoist the horses out, and swim them on shore. The risk of drowning or laming the horses by this mode must be apparent. Admitting it to be possible to swim horses on shore under such circumstances, I conceive it would not be practicable to embark horses in that manner; for, in the first place, it would be most difficult to get them to take the water through the surf; and even should they be got off, there would be still a great difficulty in slinging them to hoist them in. I am informed that in the retreat from Corunna in 1809, there were a number of valuable horses in the town the night before the battle; and although there were transports in the bay, yet, from the want of means to embark them, the whole were destroyed.

It may also be stated that the flat-bottomed boat is considered a great evil on board a ship of war or transport; it is very heavy to hoist in or out, and it occupies a large space on a ship's deck; or should they be placed on skids over the quarter-deck, they are, from their elevated position, a great encumbrance.

In order to overcome these difficulties, I constructed a model, of which the following is a description.

My first object was to devise a portable boat, one that might be easily taken to pieces, and stowed away in a comparatively small space, and which might be put together, not only with facility, but by any description of men. During the time I had the superintendence of the building of the boats for Captains Franklin and Parry's northern expeditions, I had frequent opportunities of observing Macintosh's patent water-proof canvas; one boat was built and covered with it in this dock-yard,* and one by Colonel Pasley, Royal Engineers, at Chatham, for Captain Franklin's overland expedition; and I was strongly impressed with the idea, that a canvas of this description, but much stronger, and finished in a different way, might materially assist me in making the boat portable.

The form of the boat became the next consideration. It was desirable that it should be such as would be best calculated to encounter a surf, and not longer or broader than to contain four horses (which number appeared sufficient to land at a time), and to be of such a depth as to preclude the possibility of the horses spring-

* Captain Franklin did not take this boat.

ing out. I am sanguine in the belief that I have been successful in the form, as it respects encountering a surf, as I am told she resembles the Massulla boats of Madras.

In order to make the boat portable, and to do it in as simple a manner as possible, the different parts of her frame are connected by iron screws and brass nuts; the screws are so made that they may be turned in by hand, consequently no tools are required, and she may be put together by any men of common understanding. To facilitate which, the four quarters of the frame of the boat are painted of different colours, and marked; therefore it is almost impossible to err; the screws also are all cut with the same die, and the only precaution necessary is with regard to their length. The boat is constructed with a flat floor, and there is a strong platform for the horses to stand on, which is carried sufficiently high to prevent the possibility of an accident; there are three thwarts at the proper height, for the security of the boat, and parting bars are placed on the gunwales, to keep the horses separate and steady. When the frame of the boat is put together, a waterproof canvas covering is put over it, and laced to the gunwales, when she becomes a perfect boat.

As it is intended to draw the boat up and down a beach, it becomes necessary to protect that part of the canvas which might come in contact with stones or uneven ground; therefore she is placed on a sledge, which will keep her nearly a foot from the ground, and also perfectly upright and steady when hauled up.

The boat is twenty-four feet long, eight feet six inches broad, and four feet deep, and when taken to pieces, it may be packed (with the exception of the keel and gunwales) in two cases, each ten feet long, two feet six inches broad, and two feet deep; and is not only competent to carry four horses, but also their harness, and whatever stores may be necessary.

The bulk of these cases was considered to be, and is objectionable; therefore in accordance to a suggestion made at page 549 of the Number for November 1829, of the *United Service Journal*, I have made the thwarts and part of the frame into cases for containing the remaining part of the boat, so that not any wood, iron, or rope is required in making the cases but what is indispensable in forming the boat; and the whole is now in two packages, one thirteen feet six inches long, four feet broad, and four feet deep; and the other eight feet eight inches long, two feet eight inches broad, and two feet six inches deep.

When it is required to land horses, the boat is to be put together on the ship's deck, and hoisted out; the horses must be lowered into her, and secured in the usual way. The boat is then transported as near to the beach as the surf or local circumstances will admit of; a rope, previously fastened to the sledge, is got on shore, and a sufficient number of men will haul her through the surf and half her length up the beach: the canvas at the fore part of the boat is then unlaced and rolled under the bows, the screws at the scarphs of the gunwales, at the risings, footwaling, and keel (ten in number), are taken out, when the fore part of the boat may be removed, and the horses can walk out.

When it is required to embark horses, the boat is put together on the beach, the horses led into her, the bow closed up, and with the assistance of an anchor or boat, moored without the surf, she is heaved afloat.

On the 26th of November, 1827, I had the honour to submit and to explain my model to the Honourable Navy Board, who declared themselves perfectly satisfied with it, and ordered a boat to be built on the above principle in this Dock-yard, which was done, and examined by His Royal Highness the Lord High Admiral, as well as by many distinguished Officers of the Navy and Army, and it was subsequently sent to Portsmouth for trial on Southsea beach; and on the 29th of August, 1829, the boat was put together in Portsmouth Dock-yard, when four horses were lowered into her, whilst in the Camber; she was towed out of the harbour to midway between Southsea Castle and the bathing-houses, and about forty fathoms from the shore. A rope, previously made fast to the sledge, and thirty men, were landed from the other boats, when she was hauled on shore, the canvas unlaced, the bow removed, and the horses walked out without wetting their hoofs, the whole being done in six minutes. The horses were re-embarked and got afloat in seven minutes, and returned to the Dock-yard. This experiment was made in the presence of a Committee of Admirals and Captains, and also the *Governor and* Staff of the Garrison, who expressed themselves perfectly satisfied with it.

From various experiments which I have made on the strength and water-proof qualities of the canvas used for the covering of this boat,* I feel justified in stating that it may, and that it will be used for many important and highly useful purposes, particularly as connected with the naval and military services. It occurred at the battle of Navarino, and also in many actions during the last war, that, at the conclusion of the engagement, the boats, when required to board the prizes, were found to be so perforated with shot, that they would not swim, and a considerable time elapsed before they could be made effective; and I have heard that many prizes have been lost in consequence of this

* This manufacture being composed of two pieces of cloth, united by a solution of Indian rubber, is of immense strength. The canvas used for this boat was prepared by Mr. Thomas Hancock, who manufactures this article (for Messrs. Macintosh and Co) in London, and was quite a specimen of the perfection to which this mode of water-proofing has been brought.

state of the boats. This was distinctly the case in the defeat of the combined French and Italian squadrons near Lissa, in March, 1811, by the squadron under the command of the late Sir William Hoste. Sir William states,*—"I must now account for the Flora's getting away after having struck her colours. At the time I was engaged with that ship, the Bellona was raking us, and when she struck I had no boat that could possibly take possession of her." Also at the capture of the Rivoli by the Victorious, in the Adriatic, in April, 1812, the latter ship's boats were so much injured, that they would not swim, and the prize was boarded by Lieutenant (now Captain) Peake, and one man, in a dingy.

Now if every ship of war was supplied with a covering of this description, previously fitted to one or more of her boats, which covering might be kept below until wanted, however much the boat may be hit with shot, or staved (provided her form was not entirely destroyed), by placing this covering over her, she becomes in less than two minutes an effective boat, and fit for any service.

* Vide *London Gazette*, May 18, 1811.

In obedience to verbal directions from His Royal Highness the Lord High Admiral, a small boat was built in this Dock-yard, on the principle of the horse boat: she was twenty feet long, five feet ten inches broad, and two feet two inches deep, weighed five hundred weight two quarters, and was packed in two cases, one fifteen feet long, fourteen inches broad, and nine inches deep; the other six feet one inch long, one foot ten inches broad, and one foot four inches deep. This boat was taken in the yacht on His Royal Highness's visit to Portsmouth and Plymouth in July 1828, and was put together at each of those ports, when thirty men were embarked, and she was rowed round Plymouth harbour with that number of men. Not the least weakness was apparent, and not a drop of water came into her. She was subsequently sent to Portsmouth for His Majesty's ship Madagascar, at the request of Captain the Honourable Sir Robert Spencer. Also, two boats, similar to the one supplied to the yacht, have been built for the settlement at the Swan River, Western Australia.

For military purposes I feel confident that

portable flying bridges, on the principle of those at Antwerp, and at several towns on the Rhine, might be constructed in this manner with great advantage, and when made light, and suitable to pack in a small space, might be eminently useful to an army.

Perhaps I ought to apologize for presuming to speak on military subjects; and I must again observe, that I pretend to no other knowledge than what I have collected in conversation with experienced officers, and from reading. In Sir Walter Scott's History of Buonaparte, there are many passages which strengthen my statement, *viz.*, that circumstances have arisen where boats constructed in the manner above described would be exceedingly useful. In vol. iii., p. 119, Sir Walter states, that "Buonaparte himself observed that the passage of a great river is one of the most critical operations in modern war." Page 122.—"They had to pass (the Po at Placenza) in common ferry-boats, and the crossing of the advanced guard (five hundred men) required nearly half an hour; so that the difficulty, or rather impossibility of achieving the operation, had they been seriously opposed, appears to demonstration."

"The vanguard having thus opened the passage, the other divisions of the army (between fifty and sixty thousand men) were enabled to cross in succession; and in the course of two days, the whole were in the Milanese territory, and on the left bank of the Po."

But perhaps the most striking case is mentioned in Volume vii., p. 212, when treating on the advance of the French army towards Moscow:—"The river Willia being swollen with rain, and the bridges destroyed, the Emperor, impatient of the obstacle, commanded a body of Polish cavalry to cross by swimming. They did not hesitate to dash into the river, but ere they reached the middle of the stream, the irresistible torrent broke their ranks, and they were swept down and lost almost to a man."

On the Manner of Fitting Carronades in Boats, and Suggestions on the Advantages of Mounting a Gun instead of a Carronade.

IT was observed to me by Rear-Admiral Hollis, to whom I had the honour of shewing and explaining my models for the anchor and guns, that it would be very desirable if some person would turn his attention to the way in which carronades were fitted in boats, as by the present method, should the boat be caught in a strong wind, it would be very difficult, and often impracticable, to get them from their elevated position.

The manner of fitting slides for carronades in boats has been frequently changed; formerly the slide ran the whole length of the boat, so that the carronade might be removed on it, and fired either forward or aft. At the latter part of the last war, the slide was ordered to be fitted to extend from the bow to the third or fourth thwart from forward. This slide was cut asunder about one-third from the fore end, and a hinge attached to it; by unshipping the third or fourth thwart, the after-part of the slide was lowered down to the bottom of the boat and formed an inclined plane, so that when the carronade was not in use, it was slid down this inclined plane, and consequently the top weight was in a great measure removed; also, this slide was to be fitted in such a manner that it might be removed to the after-part of the boat, should it be necessary to fight the carronade there. Subsequently, by order of the Admiralty of the 11th of August, 1823, it was directed "that the boats of line-of-battle ships be established with two eighteen-pounder carronades on slides, fitted to traverse on their centre, for launches; and one twelve-pounder carronade on a similar slide, for barges and pinnaces."

I do not know of any directions having been given as to the manner in which these boats were to be fitted.

The fitting of slides to traverse on their centre has been frequently done, but it became necessary that some provision should be made to remove the carronade to the bottom of the boat with facility and perfect safety—likewise to remount it with equal facility, even should there be considerable motion in the boat; and it would be most desirable, could it be effected without unshipping the thwarts, or disturbing the men at the oars. This cannot be accomplished by an inclined plane from a slide fitted in any way, and particularly from one which traverses on its centre.

Impressed with the observations which the Rear-Admiral had made, and in order to overcome these difficulties, and also to endeavour to carry into effect the orders received, I was led to construct a model, of which the following is a description.

The third thwart from forward was considered best for the slide to traverse on, and it was brought so far aft for this reason:—should it be necessary to fire the carronade over the broadside, it would be in the better bearing of the boat than if further forward, and consequently the recoil would not affect her so much; but by placing the carronade so far aft, it will be apparent, that when it was required to fire directly over the stem, the explosion would be dangerous to the bows: therefore, to obviate that objection, a short piece of slide was attached by hooks to the fore-end of the traversing slide, and the usual pivot-bolt was placed in the fore part of it; the traversing pivot-bolt was then removed, and it became a continued slide, similar to the former method of fitting them, and the carronade might be run close forward: it would then be optional to use the old method, or the traversing one, as circumstances might require.

To lower and remount the carronade required more consideration, and I proposed to accomplish it as follows:—

A square frame, sufficiently strong to bear the weight of the carronade, was placed under the thwart, on which the after-part of the traversing slide rested, and the thwart abaft it. Four bolts, about one inch and a half in diameter (which I name guide-bolts), one at each corner, are passed through the thwarts and frame, and into the foot-waling; up and down these bolts this frame slides. Two gun-metal screws, from two to three inches in diameter, according to the weight of the carronade, are placed in midships (and in a direct fore and aft line) through the thwarts and through the frame, and into an iron step on the keelson. A cast-iron nut is fixed in the frame, the upper part of the screws are flush with the upper part of the thwart, having a square hole in each to receive winches; on this frame a piece of plank, about fifteen inches broad, and the thickness of the thwart, is placed athwartships, midway between the thwarts, and bolted to the frame with forelocked bolts.

The carronade is then put on the traversing slide, which is placed in a fore and aft direction; the short piece of slide before mentioned is removed from the fore to the after part of the traversing slide, and connected to it with hooks as before. A pivot-bolt is placed through this short piece of slide, and securely forelocked underneath the plank, which is placed on the frame; the carronade is moved aft to this short piece of slide, which is then disengaged from the traversing one, and together with the carronade is turned athwart the boat. The winches are placed in the upper part of the screws, and it is lowered down by these screws, so that the upper part of the carronade is nearly level with the upper part of the thwarts, the guide-bolts keeping it perfectly steady, and rendering it impossible to get out of its place.

The carronade is mounted in precisely the same manner.

The Honourable Navy Board has granted permission for the launches of the Ganges, Barham, Hussar, and Briton, and also the pinnaces of the Menai, Eden, Rattlesnake, Slaney, Grasshopper, Kangaroo, Pearl, and Samarang, to be fitted according to this method; and an eighteen-pounder carronade has been removed from the fore-part of the boat, and lowered to the level of the thwarts by four men, in less than two minutes; and likewise remounted, and placed in a position for firing, by the same number of men, in the same space of time, without disturbing or unshipping any of the thwarts.

The advantage contemplated by this method is, that should the boat be caught in bad weather, it enables the carronade to be placed where it becomes as ballast, and therefore a positive benefit. Also the carronade is kept down till wanted; and it does not at any time prevent the oars from being used.

It is considered that a launch built according to the new principle is fully equal, both in size and strength, to bear the firing of a larger gun than is at present established; *viz.*, eighteen-pounder carronades for line-of-battle ships, and twelve-pounders for frigates: and that the objection to having heavier metal arises from the weight being too great to be handled in a boat.

It is also the opinion of every officer I have conversed with, that a gun would be more effective than a carronade, provided means could be adopted to deaden the recoil occasioned by the greater charge of powder required for a gun; and also facilitate the great weight being taken from the height in which it is necessarily placed, and moved sufficiently low with safety and expedition.

The launch built for the Ganges is, according to the opinion of the late Commissioner Sir George Grey, equal to a thirty-two-pounder carronade. A brass twelve-pounder gun is about the same weight; and it is suggested whether it would not be desirable for every line-of-battle ship to be supplied with a gun of that description, with its field-carriage complete, but without the limbers; also a carriage for the gun adapted to the slide in the launch, which slide might be so fitted, that when the gun is fired it should recoil dead on its brackets; and when it is to be run out, the trucks might be brought into action to lessen the difficulty of performing it; also it might be lowered from the slide by the same process as before described for the carronade.

The advantages presumed to be gained by this arrangement are a better armanent for the launch; and also, should it be required to carry on operations on shore, there is a most effective gun for that service, which may be landed and re-embarked with safety through a surf, and worked with as much celerity as can ever be expected in such operations.

It is also submitted that, in the event of another war, the having a field-gun, which may be landed and re-embarked with facility, is likely to become an object of much greater importance to our Navy than formerly; for since the introduction of chain cables, the difficulty of cutting-out vessels will be very much increased, so that means will no doubt be adopted to endeavour to destroy or drive them from their anchorage.*

* Field guns are now supplied to the Navy.

On Loading a Boat with Water from a Beach on which there is a great Surf.

Feeling anxious that the windlass in the boat should be made applicable to as many purposes as possible, and having heard, and readily conceiving the great difficulty that must arise in loading a boat with water from a beach on which there is a great surf, I was induced to consider that cases may occur when it would be desirable to use the windlass in order to faciliate the accomplishment of that object.

I am informed that the methods at present resorted to are as follows :—Should the well or spring, from whence the water is to be procured, be at any distance from the beach, the casks are filled, and rolled to the water's side; they are then separately hauled off to the boat, and hoisted in, to effect which a pair of sheers, or triangle, is erected on the ground, or a stump is placed in the mast step, from which a tackle is suspended. Or should it not be practicable to hoist the casks into the boat, it is usual to form them into a raft, and tow them off to the ship. It can easily be imagined that the distance to the ship, the wind or current, may make that a work of great difficulty and loss of time; also, that the men, in forming the raft, invariably get wet, which, particularly in tropical climates, is a circumstance of serious importance, oftentimes materially affecting the health of the seamen employed.

The present proposition is to anchor the boat without the surf, with a heavy kedge-anchor; to raft the casks as near the water's edge as possible; to get the raft rope over the stern, reeve it through a leading block, which is hooked under, and bring it to the windlass in the boat; by so doing you get a purchase equal to the heaving off a raft of casks sufficient to load the boat. When it is off, the casks should be singly disengaged from the raft rope, and taken under, and parbuckled up the stern, by the parbuckling ropes being brought to the windlass. Two skids are placed up and down the after-part of the transom, to facilitate its being accomplished; also, a strong piece of wood is placed across the boat at the height of the upper part of the transom, and about eighteen inches before it, to receive the bilge of the cask. The full cask then remains elevated, and resting against the transom and the piece of wood that is placed across; and in order to avoid the difficult and hazardous operation of moving the full cask, an empty one is placed quite forward, which should be filled by a hose from the one which was parbuckled over the transom. Its being placed so much higher will preclude any obstacle in performing it; and when empty, it is to be removed to the side of the one just filled, and proceed as before, until every cask the boat can stow is filled, except one, which of course remains empty.

I am fully aware that this mode will occupy considerably more time; but my hopes are that the object may be accomplished by this method when it is impossible to do it by any other; and should it never be necessary, or prove practicable, there is not any loss, as nothing more is required than what is to be found in the boat.

It was intimated to me by a naval friend, and I have since proved it by experiment, that instead of rafting casks with their bungs uppermost, as is invariably the practice, the better way would be to place their bungs downwards, for should there be any defect in the bung, or bung stave, so as to admit the salt water, it will of course find its way into the cask should the bung be uppermost; on the contrary, if it be downwards, the pressure of the fresh water in the cask will entirely preclude the admission of salt water; and although the fresh water may leak out whilst being hoisted in, still what remains will be pure, and not rendered brackish by the mixture of salt water: the bung is also less liable to be started out.

It now only remains for me again to crave the indulgence of those officers who will do me the honour to peruse these remarks, and to assure them, that my object and desire is to do everything in my power that may afford a chance of being, even in the smallest degree, useful to that branch of the service to which I am attached.

NOTE.—It will be perceived, on referring to the *United Service Journal* for the month of November, 1839, that Macintosh's India-rubber Cloth, alluded to at page 66, has been employed with success in the construction of Pontoons, of a very portable description, subsequent to the publication of the first Edition of this Volume.

Copies and Extracts from Official and Private Reports.

(Copy.) *H. M. Ship, Seringapatam,*
Lisbon, 31*st July*, 1821.

GENTLEMEN,

In compliance with your request, contained in your letter of the 28th May, I have to acquaint you that I have made the undermentioned experiments of the launch fitted on Mr. Cow's plan, *viz.*—

		foot.	inches.
The launch, when light, draws	forward,	1	0
	aft,	1	10

Yesterday we moored ship, hove short on best bower, and suspended the small bower anchor from the boat's bottom, without the smallest difficulty, with a kedge laid out, and two boats at intermediate distances, holding up bights of the cable, the anchor was hauled out by the people in the launch, and let go from the bottom, the ship being thus moored as before.

		feet.	inches.
Draught of water with the anchor suspended	forw.	1	6
	aft,	2	6

This day we put a bower cable into the fore part of the launch, and suspended the bower anchor from her bottom.

		feet.	inches.
The draught of water was then . . .	forward,	3	0
	aft,	2	7

We then put another bower cable in, with the principal part of it aft.

		feet.	inches.
The launch then drew	forward,	2	11
	aft,	2	10
Upper part of wash-streak from water-line . . .		1	6
Water within the trunks from the top		0	1

With this weight two pinnaces towed her at the rate of one knot against a flood-tide.

On the small bower anchor being let go from the launch yesterday, a double buoy-rope was left fast to the crown of the anchor, which being taken through the trunks, and brought to the windlass, with jiggers to the ends of the levers, the anchor was weighed, and hove up under the boat's bottom with great ease.

From these experiments, I trust the Board will agree with me as to the great utility of this plan, and the benefit it is likely to prove to the Naval Service, by affording the means of saving a ship from wreck, when no other means could be obtained.

As you will perceive that with the weight of the anchor and two cables, the water was only one inch from the top of the trunks, I should beg to suggest that leather hoses be nailed on the top of the skid to which the trunks are fixed, of sufficient length to come up just clear of the under part of the windlass, that, after the operation of suspending the anchor, a seizing may be clapped on to prevent the water washing over them; and as there is a great strain on the windlass when heaving the anchor up, I should recommend that broad iron hoops be fitted round the windlass, where the levers are shipped in.

I am, Gentlemen,

Your obedient Servant,

(Signed) SAMUEL WARREN, Captain.

The Principal Officers and Commissioners of H.M. Navy.

(Copy.) *Tony-burn, Fifeshire, Nov.* 26, 1821.

SIR,

As a projector, you will no doubt be anxious to know how one of your's answers, which I have had an opportunity of seeing tried much to my satisfaction.

It is the way you have fitted boats for weighing and carrying out anchors, as the Sybille's was about fourteen months since. Whilst First Lieutenant of her, and we were riding near Vera Cruz, Gulf of Mexico, we weighed the small bower anchor, after being down three weeks, during which time we had a heavy gale for two days, which was done with great ease by twelve men, and again laid out, by one boat towing the launch; the ground was muddy, and the anchor had good hold.

I am not aware whether Captain Rowley has made any report on it; perhaps he may require more trials, but I conceived it would be some satisfaction to yourself, from one who wishes his humble opinion could be of use in forwarding your designs for the good of the service.

I remain, yours truly,

(Signed) P. SALMOND,

Late First Lieutenant H. M. Ship Sybille.

Mr. Cow, Master Boat-Builder, Woolwich Yard.

(Copy.) *New Broad Street, May* 12, 1824.

SIR, *Ship Owner's Society.*

The Committee of this Society, having had under consideration your improved method of laying out or weighing bower anchors, have directed me to inform you that they highly approve of the plan, as being at once simple and effective.

I am, Sir,

Your very obedient humble Servant,

(Signed) S. COCK, Sec.

Mr. John Cow, H. M. Dock-Yard, Woolwich.

Present.

Lieut.-Gen.	Ramsay.	Lieut.-Col.	Williamson, C.B.
,,	Young.	,,	Drummond, C.B.
Colonel	Griffiths.	,,	Fraser.
Lieut.-Col.	Buckner, C.B.	,,	Sir A. Dickson, K.C.B.
,,	Adye, C.B.	Major	Payne.
,,	Fyers, C.B.	,,	Caddy.
,,	Sir A. Fraser, K.C.B.		

(Copy.)

SIR, *Woolwich*, 27 *July*, 1822.

In reference to your letter dated 11th March last, transmitting a communication from Mr.

John Cow, the Master Boat-Builder in the Naval-yard at Woolwich, proposing a mode for lessening the difficulty of landing guns upon a beach, and signifying the Honourable Board's desire that I should bring the project before a Committee of Colonels and Field-Officers of Artillery, in order that they might examine and report thereon for the Honourable Board's information:—

I have now the honour to report that the Colonels and Field-Officers above mentioned assembled and witnessed a trial of Mr. Cow's new mode of landing and embarking heavy ordnance on a beach in a surf, when the boat cannot approach near enough to parbuckle them out in the usual way.

The method proposed is to carry two twenty-four-pounder guns under the bottom of the boat, one on each side of the keel; to bring the boat to an anchor without the surf, and veer away cable to a prudent distance from the shore, so as to prevent the boat striking on the gun; to send ropes, previously fastened to the gun, to the people on shore, who will pull the gun to them as it is lowered from the boat.

Should the operations be carried on where there is a rise and fall of tide, and the gun should not be immediately wanted, it will be only necessary to take the boat in at high water, and drop the gun, which may be taken up as the tide leaves it; in either case the boat is instantly fit for any other service without risk of damage.

The interior of the boat is left clear for stowing the carriages and other stores.

The embarking guns is done in the same way, care being taken to have anchors of sufficient power to moor the boat without the surf, so as to counteract the weight of the gun, and the resistance it meets with in being hauled off the shore.

The boat used for trial, was the man-of-war's launch, belonging to the Royal Arsenal Repository, which, in obedience to His Grace the Master General's order, was fitted for the purpose, under Mr. Cow's inspection.

The guns were two iron twenty-four pounders, lashed separately on beds or cradles; they were lowered from the crane at the west end of the wharf, and the guns were landed at the causeway opposite the convict ship; the operation of slinging the two guns, and lowering and fastening them under the boat, took half an hour; forty-seven men were put to the rope and hauled the first gun on shore with great ease.

The second gun was tried first with twenty men, without effect; ten more were then added, and they started the gun for some yards, but it was found necessary to put ten more to draw it completely on the beach.

The guns were landed, and drawn up the beach in twenty minutes.

The first gun was hove off the beach by the boat's windlass, and it took thirty-two minutes to get it under the boat's bottom.

The second gun was hauled off by twenty-nine men, in the boat, without the windlass, and was got under the boat's bottom in twelve minutes.

The experiment appeared fully to have answered Mr. Cow's intention, and was perfectly satisfactory to the Committee.

This being the first attempt of the kind, a much longer time was taken in the operation than will be required after practice.

The Committee would suggest, that this mode should be adopted as one of the exercises of the regiment.

As Mr. Cow belongs to the Naval Department, the Committee hope they may not be considered as overstepping their line of duty, in expressing their wish that the Lords of the Admiralty may be informed of the favourable opinion they entertain of Mr. Cow's ingenuity and merit, in bringing forward an improvement to the Artillery Service.

I have the honour to be,
Your obedient Servant,
(Signed) JOHN RAMSAY,
Colonel, and Lieutenant-General Commandant.

R. H. Crew, Esq., Secretary, &c. &c. Ordnance Office.

Copy of a Letter from Lieut. Colonel Sir Richard Williams, K.C.B., Commandant Royal Marine Artillery.

Portsmouth, August 10, 1827.

SIR,

I have the satisfaction to acquaint you, that a trial was made on the 7th instant, upon Southsea beach, of landing two eighteen-pounder guns from a frigate's launch, fitted by you for that purpose, in presence of His Royal Highness the Lord High Admiral and several Naval Officers of rank.

The guns, one of which weighed thirty-six hundred-weight, and the other forty-two hundred-weight, were taken down to the beach at low water, previous to the arrival of His Royal Highness, and hoisted up under the boat's bottom by means of her windlass.

One of the gun-carriages was placed in the launch between the thwarts.

The launch was veered in towards the shore, and checked at about fifty yards from the beach, when the guns were lowered down from the bottom of the boat; and upon the presumption of a surf, and a tide that would have rendered it difficult to convey a small line to the shore, a brass $5\frac{1}{2}$-inch mortar was placed in the stern sheets, and a shell (to which a line was attached) was projected from it with a charge of two ounces of powder.

As soon as the line was ashore, the carriage was put overboard with a hauling line. The heavier gun and its carriage were hauled ashore to the distance of about fifty yards from the water's edge, by the manual labour of sixty or seventy men; the gun was immediately mounted by the operation of parbuckling, loaded, and fired.

From the time of firing the mortar, to the discharge of the gun, fourteen minutes only elapsed.

I am, Sir,

Your obedient servant,

(Signed) R. WILLIAMS,

Lieut. Col.-Commandant R. M. A.

Mr. Cow, Master Boat-Builder, Woolwich.

P. S.—This operation was under the immediate direction of Lieutenant Stevens.

(Copy.) *H. M. Ship Cyrene, at Woolwich, December* 26, 1825.

MEM.

We, the undersigned, the Commanders of His Majesty's sloops Cyrene and Beagle, beg leave to state, that in pursuance of orders from the Lords Commissioners of the Admiralty, we this day attended an experiment of Mr. Cow's plan for landing a twelve-pounder field-gun on a beach.

The gun was lowered into the water, and hove up to the bottom of a common twenty-five feet cutter, by a windlass on Mr. Cow's plan; and the whole of the tackling, magazine, and shot put into her, which, together with the gun, weighed 41 cwt. 3 qrs. 14 lbs.; she was then rowed with eight oars for nearly half a mile, and appeared by no means overladen; she then came to a grapnel about twenty yards from the beach, and the gun being let go, and the traces brought on shore, it was run up on its wheels with the greatest facility by about a dozen men.

In stating the advantages peculiar to this invention, we beg leave to mention the following:—

1st. If it were expedient, for instance, in consequence of unforeseen casualties, such as being unexpectedly exposed to breakers, or pursued by superior force, to detach the gun from the boat, that operation may be performed with perfect safety in a few seconds; whereas, by the former method it could seldom be accomplished at all, and never without imminent danger.

2nd. The boat may, at the same time, carry troops, and use her carronade without obstruction.

3rd. The landing is effected by a twenty-five feet cutter, a class of boats with which the smallest vessels of war are supplied.

4th. As the manner of suspending and detaching the gun is extremely simple, the plan is not likely to be marred by any want of practice in the boat's crew.

5th. In the event of the apparatus not being fitted in the Dock-Yard, a temporary windlass, with canvas hoses, can be fitted by the carpenters of the ship, in two hours, in a manner perfectly efficient.

There are other advantages which are, however, so obvious, that it is perhaps unnecessary to mention them. Such as, the gun is masked, and secure from the shot of the enemy, and the interior of the boat is left clear for men and stores. To this we may add, that in cases where it might be desirable to avoid all noise while effecting a landing by night, lest an alarm should be given to an enemy in the neighbourhood, this method admits of the stillness and despatch most likely to ensure secrecy.

It is therefore our decided opinion, that Mr. Cow's plan will be found of the greatest utility in landing and shipping guns of all descriptions;

and we have much pleasure in expressing our opinion of the great value of this new application.

(Signed) A. CAMPBELL,
Commander of H. M Sloop Cyrene.

PRINGLE STOKES,
Commander of H. M. Sloop Beagle.

To John Wilson Croker, Esq. &c., Admiralty, London.

Present

Lieut.-General Cuppage.	Colonel Sir A. Fraser, K.C.B.
Major General Miller.	„ Sir A. Dickson, K.C.B.

(Copy.) *Woolwich*, 26 *Dec.* 1825.

MY LORD DUKE,

I have the honour to report, that in obedience to your Grace's directions contained in Lord Fitzroy Somerset's letter, of the 23rd instant, the Officers named in the margin attended this day at an experiment made by Mr. Cow, in presence of Captains Campbell and Stokes of the Royal Navy.

Mr. Cow's experiment consisted in a mode of landing a twelve-pounder field-piece mounted on its carriage, and slung under a boat fitted with slings and windlass in the same manner as submitted some time ago for landing heavy ordnance, and the mode appears to the Committee to be well adapted to the purposes of landing ordnance from boats fitted upon Mr. Cow's principle, by which boats are capable of conveying much greater weights than they could carry in the usual manner, and much facility is effected in landing field-pieces in the surf, on an enemy's coast.

I have the honour to be,

&c. &c. &c.

(Signed) R. DOUGLAS,

Director General.

To Field Marshal His Grace the Duke of Wellington.

Extract from the Report of the Committee who witnessed the Experiment of Landing and Embarking Horses on Southsea Beach.

"29*th August*, 1829.

"The horses were lowered into the boat at the Dock-Yard, disembarked at Southsea beach by taking out the bow; re-embarked, and the bow replaced with great facility.

"The Committee is of opinion that the boat appears to answer the intended purpose, and that she may be also advantageously appropriated to embarking and landing troops, baggage, &c.

"She was put together with ease, proved perfectly water-tight, and when packed in cases, may be conveyed in vessels incapable of carrying flat boats."

Copy of a letter from Commodore White.

Woolwich, 28*th June*, 1827.

Sir,

Your method of fitting launches and other boats, for laying out and weighing bower anchors, and for landing and bringing off heavy guns and field-pieces, is, in my opinion, deserving of every encouragement, and will prove very useful to His Majesty's Service.

I am, Sir,

Your obedient humble Servant,

John C. White.

To Mr. Cow.

Dock-Yard, Feb. 22*nd*, 1838.

Dear Sir,

I have read your book with pleasure and instruction, and have no doubt of the utility of your inventions. I beg to return the enclosed papers, and hope to be allowed to keep the book.

I remain, dear Sir,

Yours faithfully,

(Signed) Phipps Hornby.

Mr. Cow, Dock-Yard, Woolwich.

REFERENCES TO THE FIGURES.

PLATE I.

Figures 1 *and* 2.—View of a Seventy-four Gun Ship's Launch, with a bower anchor suspended under the bottom, and a bower cable coiled into the boat.

a. the cable; *b. b.* the anchor; *c. c.* the buoy rope; *d. d. d.* rope by which the anchor is hove up or suspended, when carrying by the windlass; *e.* line of floatation when the boat is light; *f.* line of floatation should a bower anchor be suspended over the stern of the boat; *g.* line of floatation when the bower anchor is suspended (as shown in *Figure* 1); a bower cable and twenty men in the boat.

Figures 3 *and* 4.—An Elevation and Midship Section of the Launch, with two thirty-two pounder iron guns suspended from the windlass.

h. h. h. the guns lashed on wood slides; *i. i. i.* the heaving-up ropes, by which the guns are hove up to the boat's bottom; *k. k.* water-tight removable wood trunks, through which the ropes pass to the windlass; *l. l.* the windlass in two parts, connected by a wrought-iron gudgeon and socket; *m.* a strong removable wrought-iron stanchion, which supports the middle of the windlass; *o.* the line of floatation, when there are two thirty-two pounder guns suspended, their carriages, ammunition, stores, and twenty men in the boat; *p.* the rope by which the gun is hauled on shore when landing, and off the shore when embarking; *q. q.* guy lines, to steady the guns when under the boat's bottom.

Figures 5 *and* 6.—Plan and Profile Section. *l. l.* windlass; *m.* iron stanchion; *n. n.* iron stays attached to the stanchion to keep it steady in a fore and aft direction; *k.* water-tight trunks.

PLATE II.

Figures 1 *and* 2.—Views of a Twenty-five feet Cutter, with a twelve-pounder field-gun suspended from her bottom.

Figure 3.—View of the Boat when anchored without the Surf, the gun lowered to the ground, and being hauled up the beach.

PLATE III.

Figure 1.—View of the Horse-boat afloat, with four horses on board.

Figure 2.—View of the Horse-boat hauled up the beach, the fore-part of the boat taken away, and the horses being led out.

PLATE IV.

Figures 1 *and* 2.—Elevation and Plan of a Seventy-four Gun Ship's Launch, with a twelve-pounder brass gun, mounted forward, and an eighteen-pounder carronade aft; also showing their position, when lowered into the bottom of the boat.

Figure 3.—An elevated View of a Seventy-four Gun Ship's Launch, with a butt of water parbuckled up the stern, and an empty butt placed in the fore-part of the boat, and being filled by a hose from the one on the transom.

THE END.

London: Printed by W. Clowes & Sons, 14, Charing Cross.

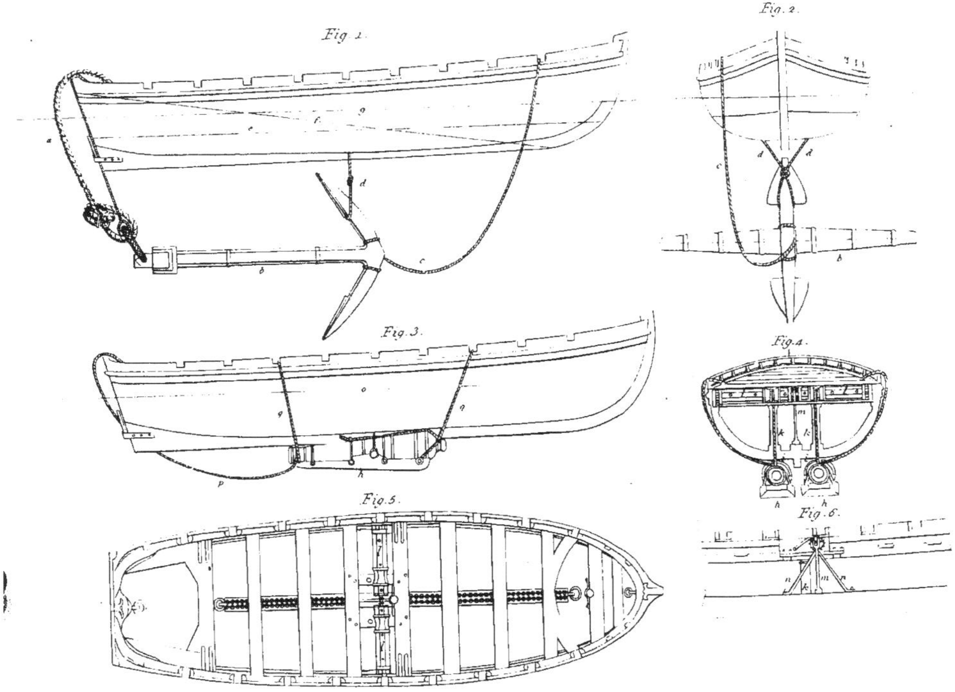
Fig. 1.
Fig. 2.
Fig. 3.
Fig. 4.
Fig. 5.
Fig. 6.

Plate 2

Fig. 1.

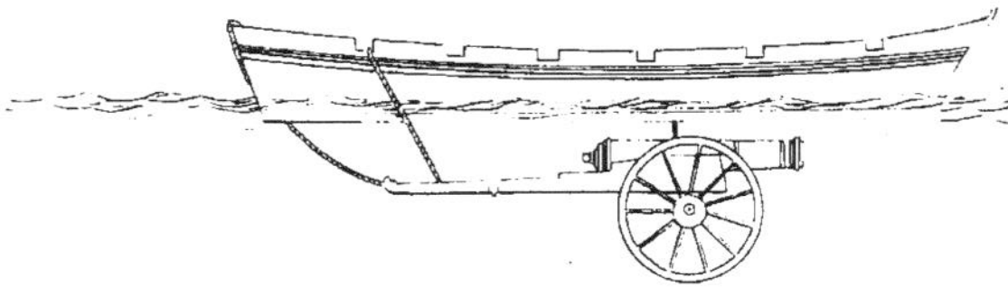

Fig. 2

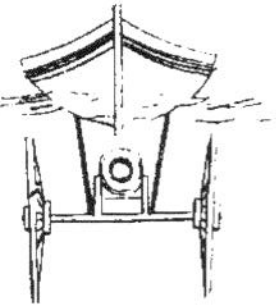

Fig. 3.

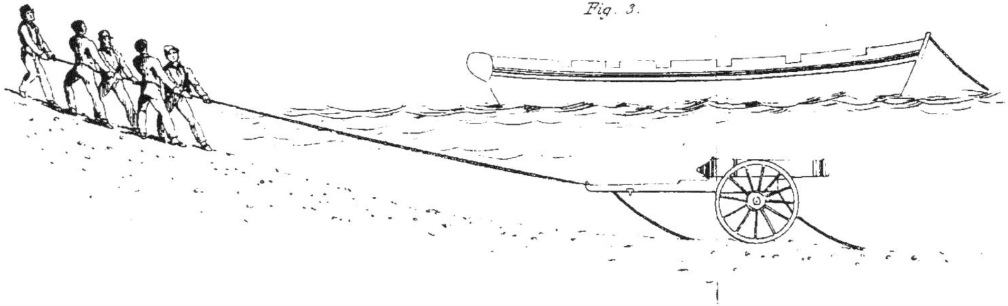

Fig. 1.

Fig. 2.

Fig. 1.

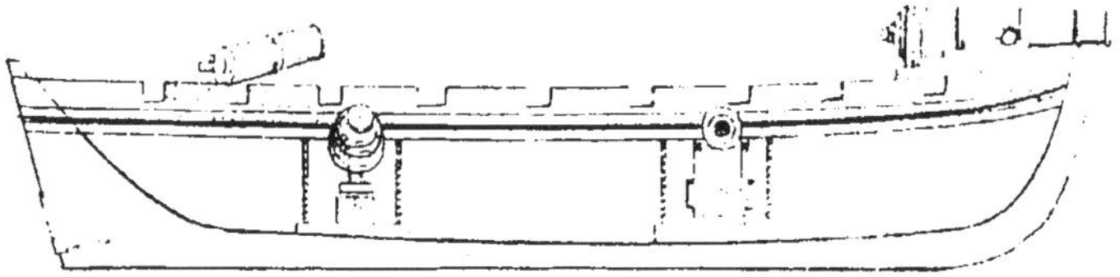

Fig. 2.

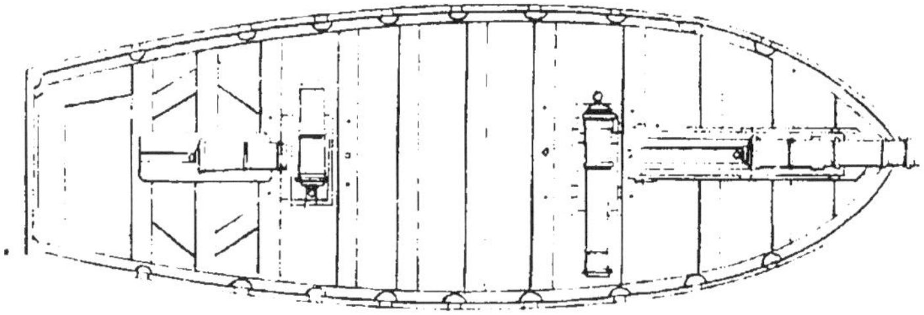

Fig. 3.

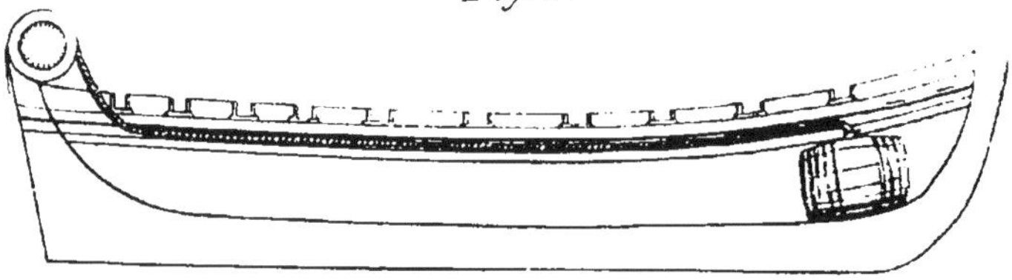

Made in United States
Troutdale, OR
11/03/2026

56844517R00062